CSI CHAPTERS

Who Knew?

By Rebecca McEwen

Contents

Ideas on the Rise

It has all the makings of a bad movie. The setting is more than 2,250 years ago in **ancient Greece**. A bunch of **philosophers** are sitting around, talking about mythology, art, and music. Then the famous mathematician Archimedes steps off an elevator and joins them.

Wait just a minute! An elevator – in ancient Greece? Back then people were just figuring out how **levers** worked. Why would anyone think that the ancient Greeks had elevators?

Well, it's because they did. Back in 230 **B.C.**, Archimedes designed a working elevator that could lift a person using ropes and pulleys. Who knew that the elevator was invented more than 2,000 years ago?

⬆ Archimedes is considered to be one of the greatest thinkers of all time.

It might seem normal to you, but an elevator was a pretty amazing idea more than 2,000 years ago!

Vending Machines

Coin slot

Take something you see every day, such as a vending machine. Supposedly, these were invented in the 1880s in England to dispense postcards and books. This is only partly correct. It was really Hero of Alexandria who invented a basic vending machine more than 2,000 years ago in order for Egyptian temples to easily dispense holy water.

Spout where the water flows out

Lever upon which the coin lands

At the time, no one really wanted to build a complicated machine to do the work that one person could do in a few minutes. Hero's vending machine was a great idea, but the world wasn't ready for it yet!

It took 2,000 years for Hero's vending machine idea to really catch on!

That's Entertainment!

Portable Media Players

Inventions are nothing new. In fact, ever since people first walked the planet, they have invented things to make life easier, safer, or more comfortable. Sometimes, though, people invent things just for convenience or entertainment.

Now, when people want to listen to music or watch any kind of video, they reach for portable media players that have massive amounts of memory and can be used everywhere. These tiny, easy-to-use machines are at the height of modern **technology** – right?

Well, if by "modern" you mean technology that has been around since the 1970s, you are absolutely correct. In 1979, Kane Kramer invented a tiny music player that was only about the size of a deck of cards and stored music as digital files on a computer chip. This machine had a display screen and earphones, and could store up to three-and-a-half minutes of music.

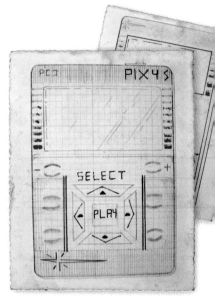

⇡ Sketches that Kane Kramer drew in 1979. Look familiar?

Sadly, this great idea didn't take off, and Kramer let the **patent** for his invention expire. Twenty years later, companies started selling portable media players, and in 2007, Apple, Inc. recognized Kramer for the invention he thought up when he was only 23 years old.

— **Fast Fact** —

In 1979, there was no Internet. If people wanted to buy a song for a fancy digital player, they had to go to a record store and download the song using a telephone line – this process took over an hour for each song! Kramer's music player truly is an example of something introduced to the public long before they were ready for it.

The Battery

An invention such as a portable media player works well today because it's tiny and can store hours of audio and video files. It also works well because it's easy to carry and, when the battery is charged, it doesn't have to be attached by a long cord to an electrical socket for power.

These batteries, which get smaller and more long-lasting every year, are another piece of technology that gets a lot of modern attention. However, most people don't know that a scientist named Alessandro Volta created the first electric battery in 1800 using common chemicals.

Alessandro Volta. The name for a single unit, or measurement, of electrical power is called a "**volt**" to honor Volta for his work with **electricity**.

In 1936, archeologists working near Baghdad, Iraq, dug up some strange clay jars that had copper-wrapped **iron** rods stuck down through the openings. These jars dated back to around 200 B.C., and may have had a special purpose. Some scientists believe these jars were designed so that, when filled with vinegar or lemon juice, they could produce electricity. Could these have been the very first batteries?

Stopper

Copper wrapping

Iron rod

Vinegar or lemon juice

Clay jar

The Computer

When you wake up in the morning, a computer tells your digital alarm clock when to ring. To heat up pizza in the microwave, you have to program the microwave's computer to say how long it should cook. Your family's car has a computer that tells the engine how to work most efficiently.

Computers play such a huge role in modern life, it's hard to believe they have only been around since Englishman Alan Turing and German Konrad Zuse built some of the earliest ones in the 1930s. These gymnasium-sized calculators were the height of technology back then, and even today Turing and Zuse often get credit for being the fathers of the computer age.

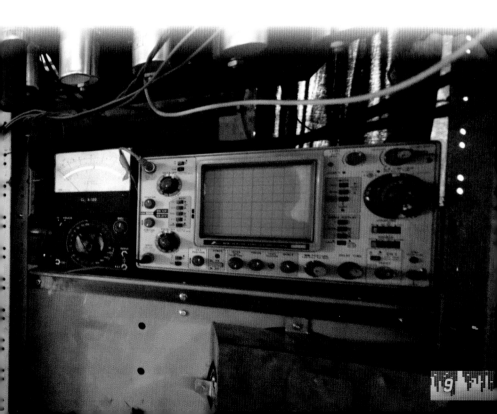

This isn't really fair, though. More than 100 years earlier, an English mathematician named Charles Babbage started inventing an engine that could compute just about any math problem. He built **prototypes** of this very early computer, but never got a chance to build the machine he designed.

In 1985, the London Science Museum started to build the computer that Babbage designed. It took six years, and used nearly 4,000 parts, but in the end, it worked exactly as Babbage had hoped it would. If Charles Babbage had been able to build his engine, the computer age might have started 100 years earlier than it did!

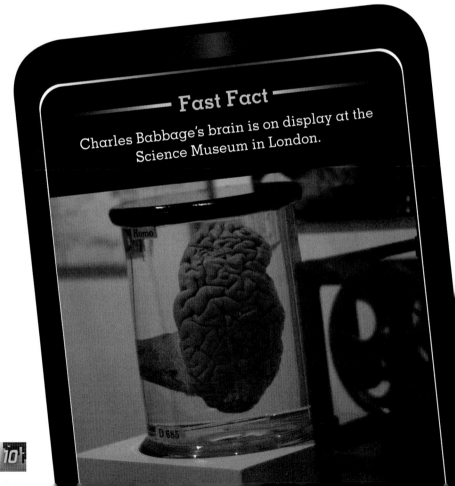

Fast Fact

Charles Babbage's brain is on display at the Science Museum in London.

It's fun to figure out where words first came from. Have you ever wondered why the word "bug" is used in computer language? Today, "debugging" means cleaning a computer of any viruses or bad data. Back in the days when computers were the size of small houses, they were powered with vacuum tubes. Designers built long walkways behind the tubes so they could replace any that burned out. The tubes gave off so much heat when the computer was running that any bugs crawling or flying nearby would die and fall into the machine. Technicians had to clean all of the dead bugs out of the circuits, or "debug" the machine.

The Video Game

Is it possible to talk about inventions and entertainment without mentioning video games? It all started back in 1972, when people were first introduced to an addictive digital ping-pong game called Pong. This game had a dark screen, and two players swiveled dials back and forth to move paddles up and down the screen and "hit" a ball across to the other player's side of the screen.

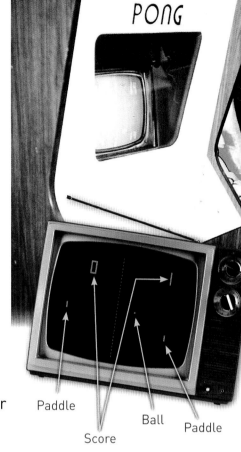

Paddle

Ball

Paddle

Score

It's tough to imagine that people ever really enjoyed playing a game like Pong when you compare it to more modern games, with their exciting graphics, complicated structures, and lifelike game controls. Today's games actually seem a lot more like the *real* first video game, which was invented in 1948. This game, which had the very boring name "the Cathode-Ray Tube Amusement Device," was based on **World War II** radar displays.

Players would turn knobs to direct light beams that served as the game's missiles and hit targets that showed on the game's screen.

Physics professor Thomas Goldsmith, Jr. created this game. He was able to get a patent for it, but it was ridiculously expensive to produce since it used real military equipment. It was never manufactured or sold, but if you want to find the earliest lifelike video game, this is it!

Going Places

The Car

Screeching around the track at 267 miles (430 kilometers) per hour, the turbocharged Bugatti Veyron Super Sport has 1,200

horsepower, and is worth over a million dollars. In 2010, it was named the fastest **production car** in the world, but it won't be long before another car will come along that beats it. This is because auto manufacturers are always trying to top each other with better, faster, and more efficient cars.

This urge to be first and best started back in the 1880s, when many talented inventors were trying to be the first to invent the automobile. In 1881, a French inventor named Gustave Trouvé built a three-wheeled car that was powered by electricity. In 1885, German Karl Benz built the first gasoline-powered car.

Still, even though Benz earned the patent, his car wasn't really the first. In about 1769, a Frenchman named Nicolas-Joseph Cugnot built a steam-powered car. Developed for the French military, this car had an extremely heavy boiler and engine on the front of its three-wheeled body, and could carry up to 4 tons of **artillery**.

↑ Cugnot's steam-powered car

The French military wasn't terribly impressed with this vehicle, because it had a tendency to tip over – quite dangerous when it had a steaming hot boiler and was surrounded by weapons. The military abandoned the project, and the world had to wait another hundred years for its first working car.

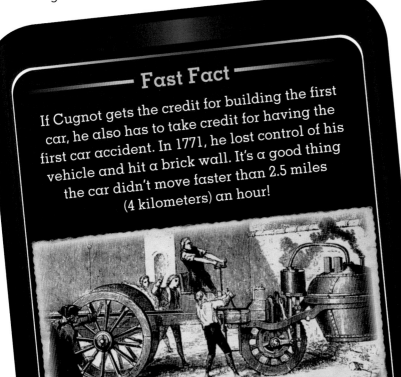

Fast Fact

If Cugnot gets the credit for building the first car, he also has to take credit for having the first car accident. In 1771, he lost control of his vehicle and hit a brick wall. It's a good thing the car didn't move faster than 2.5 miles (4 kilometers) an hour!

The Helicopter

By the early 1900s, inventors had figured out how to build working cars, even though the cars didn't work all the time. Their next goal was to invent machines that helped people travel through the sky.

Americans Orville and Wilbur Wright recorded the first successful airplane flight in 1903. Four years later, a French bicycle maker named Paul Cornu designed the first helicopter. His machine hopped along the ground, and rose no more than 7 feet (2 meters) in the air. Unfortunately, the controls on this machine didn't really work, and eventually, Cornu abandoned his design.

↑ Paul Cornu's helicopter hopped, rather than flew.

What is strange is that Cornu gets credit for building the first helicopter, even though it didn't work. If a helicopter doesn't have to work to count in the race to be first, perhaps Leonardo da Vinci should be the winner. In 1493, da Vinci designed a helicopter that he called an "air screw."

Leonardo da Vinci (1452–1519)

Da Vinci's air screw

Although he never built a full-sized air screw, da Vinci's notes show that the models he made of the craft flew. Even though his early helicopter would have never been able to lift a person into the air, da Vinci still came up with a way for this technology to work. His drawings show a person turning the air screw, but no person could ever turn it fast enough to get the heavy machine to lift. The air screw needed an engine, but they didn't exist back then.

Fast Fact

Although people were building fairly reliable airplanes by 1911, it took much longer to build a helicopter that worked exactly as it should. In 1924, the first helicopter that could fly just over half a mile (800 meters) was built.

The Submarine

While people's attempts to build flying machines were open to anyone who watched the sky, the struggle to build a machine that could travel under the water was much more secretive.

This is because militaries wanted to find some way to attack enemy ships without being seen. A machine that could sneak up from underwater, completely out of sight of any guards, seemed like a perfect solution.

During the **Revolutionary War** (1775–1783), a sixteen-year-old American student named David Bushnell designed the first wartime submarine. Called the *Turtle*, this tiny boat could hold only one person, and it was powered by a hand-cranked propeller. In 1776, the *Turtle* was put to service and ordered to sink a British warship that was docked in New York Harbor. Even though it failed, this was the first-ever submarine attack.

↑ The *Turtle* was powered by hand.

A modern submarine ⇥

It's War!

The Flamethrower

Germany is often credited with inventing the modern flamethrower back in 1901, but people have been creating this sort of fiery weapon for thousands of years. For example, in 400 B.C., Spartans learned how to use catapults to hurl burning balls of **pitch** and sulfur at their enemies.

Then, in A.D. 673, the people of Constantinople fought invaders by spraying them with a chemical called Greek fire. People shot Greek fire with either catapults or using a simple hand pump, and the effect was terrible. When the chemical came in contact with seawater, it would burst into flame, and ships hit with Greek fire burned to ashes.

⬆ Soldiers attacking a ship with Greek fire

The Heat Ray

If flamethrowers that spray burning chemicals make you think of horrible battles from the **Dark Ages**, a heat ray that can zap an enemy sounds like something straight out of the future.

Modern militaries are still creating state-of-the-art weapons. Heat rays are right in their first line of defense. In 2007, the U.S. Army demonstrated its new heat ray, which was designed to control crowds. The ray fires a beam of **radiation** at a person with the idea that the pain will make the person immediately stop whatever he or she is doing.

It makes sense that people will do just about anything to escape the pain of a heat ray. After all, in Greece 2,200 years ago, Archimedes, the same man who invented the elevator, supposedly invented a heat ray so intense it could burn down enemy ships.

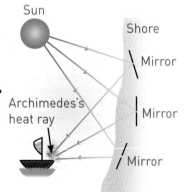

Because many modern scientists weren't convinced that Archimedes would have had the resources to create such a weapon, in 1973, a Greek scientist named Ioannis Sakkas did an experiment to try and prove this heat ray worked. Using Archimedes's drawings, he positioned 70 copper-coated mirrors so they would reflect on a plywood mockup of a Roman ship that was 160 feet (49 meters) away. Sure enough, the ship burst into flame in a matter of minutes!

Inventing for the Future

Sometimes, when thinking of inventors from long ago whose ideas are still useful today, certain names seem to come up again and again. For example, Greek mathematician Archimedes had the original vision for many inventions that continue to make life easier in the modern world.

Still, no one had more ideas than Leonardo da Vinci. The air screw was only one of his ideas. Considered one of the world's greatest artists, da Vinci also left notebooks full of ideas so far ahead of their time, it's almost as if he were predicting the future.

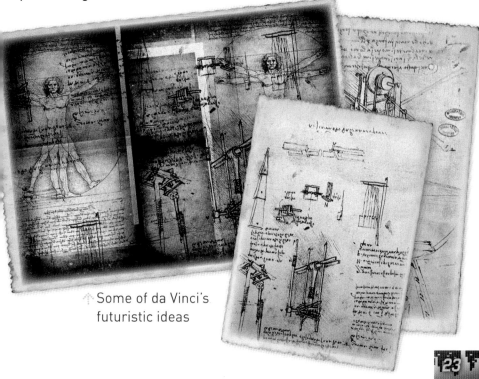

⬆ Some of da Vinci's futuristic ideas

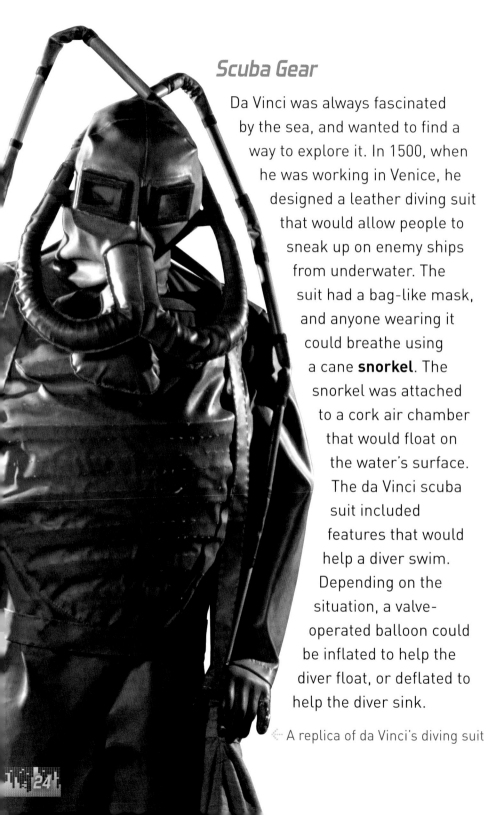

Scuba Gear

Da Vinci was always fascinated by the sea, and wanted to find a way to explore it. In 1500, when he was working in Venice, he designed a leather diving suit that would allow people to sneak up on enemy ships from underwater. The suit had a bag-like mask, and anyone wearing it could breathe using a cane **snorkel**. The snorkel was attached to a cork air chamber that would float on the water's surface. The da Vinci scuba suit included features that would help a diver swim. Depending on the situation, a valve-operated balloon could be inflated to help the diver float, or deflated to help the diver sink.

← A replica of da Vinci's diving suit

The Winged Glider

Many of da Vinci's designs and inventions were inspired by nature. Perhaps none of them demonstrated this more than his glider.

↑ Da Vinci's glider

The wings of the glider, which spanned more than 33 feet (10 meters), had a pine frame covered in raw silk, and pointy ends that looked like a bat's wings. The pilot lay down between the wings, and then made them flap by pedaling one crank and turning another. The faster the cranks turned, the faster the wings flapped.

Da Vinci never put this invention to the test. It probably would have flown, just like people fly hang gliders today.

The Self-Propelled Cart

Leonardo da Vinci created plans
for a very complicated,
small wooden vehicle
that was powered with
coiled springs. It had basic
steering and brakes. **Scholars**
and scientists tried to understand
this invention for hundreds of years.
Then, late in the 20th century, science

⬆ Model cart, based
on da Vinci's plans

was finally advanced enough for people to build the model
cart based on da Vinci's plans. Not only did the cart work,
but, strangely, it resembled another small, self-propelled
robot – the Mars Rover.

⬇ The Mars Rover

The Parachute

Although another man, Louis-Sébastien Lenormand, is considered the inventor of the parachute, it isn't actually true. Da Vinci invented the parachute several hundred years before Lenormand was even born. Da Vinci's parachute design doesn't really resemble modern parachutes, as it had a wooden frame, a coated linen skin, and a triangular shape. Throughout the centuries, many people weren't sure that his design would work, because da Vinci never got around to testing it.

However, in 2000, a daredevil named Adrian Nicholas built a da Vinci–style parachute and tried it. Not only did it work, but Nicholas insisted it gives a smoother, better ride than modern parachutes do.

So, the next time you reach for a drink from a vending machine, take a flight somewhere, get into the family car, or plug a USB cable into your laptop to recharge your phone, spare a thought for those futuristic thinkers of long ago who brought so many important and imaginative ideas into the world. Who knew? They did!

Glossary

A.D. [7] – (Anno Domini) a system used to count years in our current time. In Latin it means "In the year of our Lord"

ancient Greece [7] – a period in Greek history that lasted from about 750 B.C. to 146 B.C.

artillery – light or heavy guns and/or missile launchers

B.C. [7] – (Before Christ) a system used by some people to count years before our current time, or before the Christian (Common) Era

Dark Ages – a time in European history that started in A.D. 476 and lasted to the Renaissance, in the 1400s

electricity [3,7] – a natural form of energy

horsepower – a unit of energy equal to the energy needed to lift 550 pounds (about 250 kilograms) in one second

inventions [7] – creations

iron [7] – a metal used for making tools and machinery

levers – rigid bars used to lift things or to pry things open

patent – official protection for an idea

philosophers – people who study and think about ideas, theories, and beliefs, especially in fields such as ethics, physics, and logic

physics – the science that studies matter, energy, motion, and force

pitch – a black, sticky substance, often made from wood

production car – a car model for which many, rather than just one or a few, cars are produced

prototypes – the first versions, or models, of something

radiation – the waves of energy sent out by sources of heat or light

Revolutionary War [8] – the American Revolution

scholars – people who are very well informed about a particular subject

snorkel – a hard tube through which a swimmer can breathe while swimming facedown in the water

technology [3,6] – scientific advances that are helpful to people

volt – the international measurement for one unit of electric power

World War II [7] – a war fought from 1939 to 1945

Academic Vocabulary Key		4	Economics	8	US History	12	Technology
1	English Language Arts	5	Civics	9	World History	13	General Arts
2	Mathematics	6	Geography	10	Health	14	Dance/Music
3	Science	7	General History	11	Physical Education	15	Theater/Visual Arts